A Theory of Consciousness

By Daniel Rosenthal

The headings are not in any particular order, they include:

1 What is Consciousness? The conscious visual image.

The most obvious thing about our consciousness, something which is so obvious immediately, is that the largest part of it is a very bright and very big real image, which seems to be the whole world existing outside of us, around us, very big, going very far and being very bright, and even vividly feeling quite solid.

As we see that very big, bright, great image all around us, it is all the world to us.

It extends it seems for miles and miles around us.

I call that our "conscious visual image".

The conscious visual image is not what it looks like. It looks like the whole world existing for miles and miles around us, but actually it is our consciousness creating a bright real image of the world.

Now if your hand reaches forward and touches an object, what you see feels solid. And actually what you see is not the object, but a bright image of it.

The image which you see, now has the added additional quality of feeling solid.

I have mentioned that though it's an image it feels solid, and must somehow try to explain.

Images can be improved by adding good and useful qualities to them, qualities to make image a more accurate likeness of object.

Where people have wanted technology to improve images, they at first thought of adding color to the black and white photographs and so created color film.

And later they thought of adding movement and creating cinema films and Cinemas.

Then, people wondered if they could create an image that would look three-dimensional, and with lasers they created holograms, and some other ideas for three dimensional images. But those holograms were not all that good.

An idea has been thought of, of a cinema where perfumes would let you smell the images.

Someday someone might want an image to feel really solid, the same as the solid object it's supposed to be an image of.

Imagine that in the far future, a machine can take an accurate measurement about how solid and how heavy an object is, and then immediately create an image of the object which actually feels quite solid? So the image feels solid just like the object.

 But knowing an image like that, isn't in the future, it is all around you today, right now. Since today and right now, you have the solid feeling real images all around you.

Something inside you right now is creating really bright images which you see all around you, and they don't just look like real objects, they feel solid too. Technology can't do that now, but something natural is obviously a lot more advanced than technology, surpassing technology by a long way. Right now you see bright three dimensional images which seem solid, seem like real objects all around you.

You see a big three dimensional image of the world around you as the brightest thing in your consciousness. And yet maybe mistake it wrongly for those real objects, maybe mistake it for the solid matter of the real objects, and so not know that it's an image?

So most of consciousness is a very bright, very big, three dimensional solid looking image. I call that a "conscious visual image".

An image which seems to be the whole great world existing brightly and solidly for miles and miles around you is your consciousness and that's what consciousness is.

Now that image is consciousness. A bright real image. The image which seems to be the whole world existing for miles and miles around us, is certainly part of ourselves, it is made by something in us, yet it seems to extend for miles and miles, all far around us, and it gives us the correct vivid impression that our eyesight is creating a whole world.

As we're seeing an image of objects, that's our way of seeing objects, and we only see objects in that way by creating an image of them.

We cannot ever see real objects directly. Our sight of real objects is not and can never be direct. An image which is an accurate bright likeness of the real things forms, and we see that, brightly, as the image which seems to be the whole world.

But technically we do not ever see the real objects, we see nothing whatsoever of real objects because that wouldn't be possible. An image of the objects has to form first.

We see nothing whatsoever of real objects, though we can say loosely in speech that we are seeing the real objects because our visual images of the objects are an accurate and realistic likeness to the objects. And as information comes from our eyes about them, our bright three dimensional conscious visual image is constantly updated with more new information to remain accurate.

It vanishes when you close your eyes, showing that it was an image. It's not the real world that it seems to be, it's a very good image of all those things in the world and that is why it vanishes when we close our eyes.

How about a listing of some of our image's properties. It vanishes, disappears when you close your eyes. Its disappearing proves that it is an image. A part of the world around you would not disappear as you close your eyes. An image would, images naturally vanish and disappear. That's easy for an image.

While your enormous big, bright, image of the world is like your eyesight creating a world, you might have the impression that every single part of the image is a part of yourself. Your feeling that it is a part of yourself is absolutely correct, because it is entirely a "conscious image".

 It is obvious to all of us that the very big, bright, image is very accurately three-dimensional, with its being accurate and very well formed in three dimensional shape. A Cinema projection and a Hologram comes nowhere near to being so excellent.

And considering that our eyes are not physically so accurate, and considering that our brains are folded up with no straight lines, the accuracy of our conscious visual images is amazing.

Inside the brain you don't have straight lines. Something very skilled and very intelligent must be straightening out the distortions in the information, making the final image all straight, which takes intelligent skill.

As we look at the edge of a building wall or as we look at the straight edge of a ruler, the image we see of it is also perfectly straight, a perfectly straight line, in our conscious visual image.

We can not ever see real objects, an image has to form firstly, and then the thing we see is the image. So absolutely obviously our "conscious visual image" is the actual things we are seeing, and so obviously it is never, never, our feelings of being conscious or aware.

Consciousness itself is so obviously never, never, like our feeling of being aware, consiousness is only what seems to be the very bright three-dimensional world we see around us, as the consciousness is a bright real image.

The image looks like a bright world being created around us. It is never like something mental. It is to us a whole world being created. While reality is that what we are seeing is a really big bright real image being created, and so obviously an image of superior science, something way more advanced than a cinema projector for instance.

Our conscious visual image is always a main and most obvious part of our own consciousness, constantly mistaken perhaps for a whole world of real light and solid real objects.

It is such a big image, so bright, and so accurately three-dimensional so perfectly seriously solid in feel, that nothing in the technology of holography and cinema projection can come anywhere near it, and that's obvious I suppose to everyone.

So we have got an image all around us, which is so obviously and seriously more advanced than, and superior to, all of cinema projection technology.

A Theory of Consciousness

The image which we see constantly and continuously with our eyesight, is not like those small images that form on the retinas of our eyes. Do you wonder what converts those small images on our retinas into what we see? It's spirit, a spirit is there and it is creating a superior image, absolutely superior to the cinema and to the best hologram. It's obvious that no hologram comes anywhere near to being so good.

But why is this very big bright image not often spoken of?

Have you heard of someone speaking about seeing an enormous bright image? Has anyone said, "what an extraordinary bright image of all the world I am seeing"?

And if not, what is it that stops that?

What is it that stops him or her from ever speaking about that?

I mean what is it that would stop somebody from talking about what is the most obvious to himself about his own consciousness, and an image of such amazing qualities that its creation is a wonder and quite extraordinary?

It's clear that the stop must be the conscious image being mistaken for something different.

But what on Earth could it be mistaken for?

It might be mistaken for what it looks like.

In general when something is mistaken for something different, it is mistaken for what it looks like. And so we might assume that most people mistake it for what it looks like.

It looks like the world around us, so it can be mistaken for that.

Your image of any object in front of you, might be mistaken for nothing more than the object itself.

 Our bright image of the world, which is so good that nothing in the technology of holography and cinema projections can come anywhere near it.

 Very bright, and well formed in its perfect three dimensional accuracy. Feeling solid.

A Theory of Consciousness

We see it all day long every single day, except that it disappears at the moment when we close our eyes. And if you thought about it more sensibly you would decide that the disappearance of that bright sight is meaningful.

Its disappearance means that it was an image, as it is common for images to disappear.

The cinema projection vanishes when the projector is turned off.

The television picture disappears when it's unplugged, and when the radio wave signals carrying the picture is turned off.

The meaning of the disappearance of the whole bright world around you, as you merely close your eyes tightly, would be missed if you incorrectly thought that you are seeing the back of your eyelid.

The complete disappearance of that enormous bright image does certainly prove that it was an image.

As it is a common quality of images that they can disappear, ever so easily, while real objects do not disappear.

When you are dreaming, you often see bright images, but there is a tendency to forget about them. The brightness of images in some dreams is a clue.

What a pity that as you wake up you sometimes forget that you saw a bright real image, and I wonder if some people tell themselves they only saw a dim mental image. So have you noticed that sometimes dreams really were extremely bright real images, and not something like thoughts. And they may even feel quite solid so they seem like a world.

They were real images, real as images. So sometimes dreams can be a phenomenon of real bright real images, and utterly different from a mental phenomenon. As they were in a class of something that can be called a bright real image, and not something mental.

Which have a physical reality as something you can only call 'bright real images' like those you see when you're awake.

How do we see?

How is it that we see? The process of eyesight is that, firstly a picture with millions of pixels forms on the two curved retinas of our eyes, and then the picture goes through our optic nerves to our brains.

And then the information is transmitted to your spirit, which is a living creature which you can't see. But it's there invisibly. Each time there is an electrical impulse in any brain cell of the visual cortex area of your brain, a single bit of information is transmitted to your spirit. Your spirit gets all the information because your spirit is in contact with all the brain cells at tiny spots of contact, which should be called 'contact points'.

Your spirit then uses all that information to create a bright real image of the world that exists outside of you and around you, and this image seems to really be physical world existing around you. The visual image does not ever in the slightest extent seem to be a mental phenomenon, but instead of that your visual image seems to be made of bright light and all the physically real solid objects around you. It is therefore a 'bright real image', and while it looks like one thing it's actually something else: It is your consciousness in reality in the form of the bright real image.

In the visual cortex area of our brain, the information is at first analyzed and organized by columns of brain cells, into a form that will make it more easy for a spirit, for a different form of life which is a spirit, to interpret it and to understand what it means.

Your brain has a spirit attached to it. It's another form of life, which belongs to a different kingdom of life since it isn't made of any substance you know of. It lives in a symbiosis with the brain which is like a Lichen, since these two completely different forms of life are joined together. Spirits need information to be presented to them in a way that allows them to properly receive the information and to understand it.

And large areas of a brain are designed for that purpose, to organize and present the information so that it can be received by a form of life which can be called a spirit.

A Theory of Consciousness

This form of life exists there and it is invisible, and it is often called a spirit or a soul. And this form of life is believed to be eternal and transparent to light so it's invisible. Which is correct.

Every time there is an electrical impulse in any brain cell, a single bit of information is transmitted by that to a spirit which is in contact with the brain, and in this way billions of bits of information are transmitted all the time in parallel to the spirit, which likes and needs information.

The most common symbiosis is the Lichens which grow on damp stones. In Lichens a fungus and an algae both live together in a symbiotic association, and a spirit does the same thing with a brain.

A Spirit is a kind of invisible life which likes and needs information, information is like its food, information is what spirits need most, and any spirit's whole way of life is based on wanting and needing information. And spirits automatically try to create images which show the subjects of the information.

The images created by your spirit are immediately real to you, you see them, feel them as solid real things. They are absolutely physical, seeming to be real objects, but actually being real images. The images which you see as the world, very big, far outside of you going for miles and miles, bright with light, are the images which your spirit creates which show how much the spirit has so far been able to understand of the meaning of the billions of bits of information which it is receiving from its attachment to the whole of the brain.

Information in your brain coming from all your different senses is used to improve the images. An example of it is that, when your hand touches an object your image of the object seems to have become solid, and it's now so obviously much more like a solid model than a mere optical image.

Your conscious images are to you the actual physical objects you touch and see, the conscious images create a bright solid physical world for you, and seem to be all the absolutely real physical things around you, in front of you. They are real images.

But you do not ever see any real objects, however weakly, however faintly, you can never see a real object. Because firstly information comes about the real objects, and a very bright real image has to form, and then the image is what you see.

The conscious visual image

All day long while we are awake we all see our very bright, big, enormous, and accurately three dimensional illuminated image of the world, outside of us, enormous around us, and this image is our consciousness.

But this extremely bright real image is actually the same phenomenon as the brightest or unusually bright dreams which we sometimes have when we are deeply asleep. Created by the same mechanisms.

Of course it's not in our heads, and you don't know where it is but in impression it's so clearly a world existing far around and way out outside of us apparently.

Of course that while we are awake the bright three dimensional image is based upon true information immediately coming from our eyes, which is every second updated by more information from our eyes and other senses.

While the brighter than average dreaming image could have been based on our memories, and even random firing of brain cells, which are supplying information to your spirit which then forms images based on it.

An image formed by a lens is called a 'real image' when it can be projected onto a screen.

But when we are wide awake, our consciousness is that very bright real image, but of course it is also very obvious that your brain would not be able to create anything like the image you are seeing.

A conscious image of light is what we see whenever we look at the sun or an electric light bulb shining, and it looks so exactly like a real light that we couldn't think of any difference in our description between a real light and our conscious image of light.

But it's an amazing fact that we cannot ever see a real light, and everything we think of as light is our conscious image of it, formed from the information that comes from our eyes.

And something very incredibly advanced and scientifically superior is working to create our conscious visual image which is constantly creating a whole world for us.

Well whatever you think, no one could doubt that the image is real, the only mistake could be thinking it is something different, while the real image is wrongly mistaken for a real world of matter and objects. It is mistaken for what it most looks like. The conscious visual image is maybe mistaken for the creation of the world, because it looks like the world. You always feel that the world is created around you, and it is because you are seeing an image of it which is in fact created by you.

Weirdly books written about consciousness sometimes state that it does not exist. Most strangely, there are signs that frequently people do not understand that other human beings are conscious.

 Well caused by ignorance, many people perhaps don't believe that your eyesight is creating a whole bright world for miles and miles around you. They just might not believe that? That eyesight is creating a world. Or that you, another person see things as they do. A good example of that was a book about consciousness in which its author said and kept repeating consciousness was nothing more than a faint intangible feeling.

According to that author you can't be sure consciousness even exists in any form at all. And he was saying human beings would all be perfectly well off if they weren't conscious.

Of terminology, the obvious difference between conscious and mental.

About terminology, what is the difference between a conscious image and a mental image?

A conscious visual image is a very bright image which is at least a million times or a billion times brigher than a mental image, literally. Other types of conscious image are not visual, but they also always seem to be an absolutely real physical part of the world around you, and this includes your image of your own body, an image that seems very solid.

A mental image is always an extremely faint image, it is so faint and dark, it is intangible. A mental image does not seem solid, and is extremely dim.

When you want to form a picture of something in your head you can form an extremely faint and dim image, and you know that it's about billion times darker and dimmer than a visual image.

It is so faint, extremely faint whereas a conscious image is the opposite of faint, it is a very bright strong and obviously real image.

If you are outdoors in the sunlight, your conscious visual image is immediately about as bright as the sunlight, and so your conscious image is as bright as something lit up by at least thousands of watts of light bulbs! Of course actually you don't ever see sunlight, not at all, and what you think of as sunlight is entirely your own visual image. So pure consciousness is an extremely bright thing.

 It's as strong or stronger as the strongest cinema projector. And it's a part of you, since the whole of the outside world which you see around you is an image which is entirely a part of your consciousness.

A conscious image of sound always seems to be a real sound in the air, outside of us, physically real and loud in the air.

The conscious image of touch always seems to be the real solidity and heavy weight of objects we are touching. The conscious image of touch seems to us to be most vividly a main part of solid real objects, but it too is in reality our consciousness. Because it so vividly solidly seems to be a real solid object it is mistaken for the real objects.

Our conscious image of an object in front of us always seems to be that real object. The real object itself, never our awareness of it. The conscious image never, absolutely never seems to be our feelings of consciousness or awareness, the image only seems to be the absolutely real solid matter of the object.

So in terminology, "conscious visual image" is a very bright real image.

Whereas "mental image" is a very dark, weak, and dim faint image.

Words like perception or awareness apply exclusively to dim and weak, faint mental images, and such words cannot be used with any description of any form of conscious images.

We always have many mental images, and as well, many conscious images, all day.

The mental images we see or just feel are so extremely faint, so extremely weak that one should compare them to a scene in the dark where a flashlight has a nearly completely worn out battery.

And where the scene is so dark you barely see a thing, or don't see anything.

Now we have a whole lot of the incredibly weak mental images, which are all so extremely faint that they will not be mistaken for real objects.

And they probably have to be that weak, for an evolutionary reason, as in animals, who see and who also are conscious almost like us, the animal would in its ignorance of science believe that its conscious images are real things, a main part of a real world and real objects.

But that mistaken belief is absolutely harmless to evolution and natural selection, as it doesn't matter to natural selection whether the animal believes the conscious images it sees are real objects.

However, were the animal to believe that it's mental images are real objects, it would be a bad failure, as the animal would think it is creating things by thinking.

Rather than look for food, the animal could think about food and see it appear, and then eat what is a hallucination, as it would starve.

An animal which could just think of food and see images of food appear, would eat the imaginary food, so it would starve. Its empty stomach would not be filled. So evidently evolution would make very sure that mental images are too faint to ever be mistaken for objects. This might be the reason for the faintness of mental images?

However the conscious image based upon immediate information from eyes and senses, about the real object can be mistaken wrongly for a real object in the mind of that animal and that will never do it harm, and will never do harm to its survival.

And so evolution by natural selection will certainly allow a completely wrong mistake in your thoughts as it is so completely harmless to survival.

But differently, evolution would make sure that the mental images are so faint that you never mistake them for the creation of real objects.

So the extreme weakness faintness of mental images is selected for, while the conscious image of real objects can in evolution become a really strong and bright image, as the brightness definitely improves it.

And of terminology, that would define the difference between conscious and mental.

All thoughts are mental images which are so weak, so faint, something totally different from a conscious visual image (which is your normal eyesight,) while eyesight is creating a world so very bright and strong for miles and miles around you.

The Symbiosis of two so very different forms of life.

Two different forms of life that are so completely different in all of their physical properties and abilities, and yet they are permanently joined together in the surface layer of a brain.

How did the symbiosis start?

When two different forms of life are attached to each other, and when both forms of life gains an advantage from being attached to the other, then that is symbiosis.

And like the algae and the fungus which form Lichens.

Some Lichens actually have a folded shape, a bit like the folds of a brain. Could it be that sometimes a folded shape is typical of a symbiotic relationship between two forms of life?

As a spirit is attached to a brain, it is a form of invisible life physically attached to all of the brain cells of the brain at very many millions of tiny spots which I think are called contact points.

Of course both spirit and brain get an advantage of a form which is very easy to describe and understand, from their being attached to each other.

It's easy to see how each gets a lot of advantages from being attached to the other.

Neither the brain or the spirit would be able to do anything much on its own, it really needs to be joined to the other, and each gets really a lot from the symbiosis.

The brain is like a Lichen where symbiosis exists between algae and fungus.

The animal gets from its spirit a much higher intelligence than the brain could have without it. And it gets especially much more purposeful and skilled movements which it needs for its survival once it begins to depend on it.

A Theory of Consciousness

As an algae and a fungus can be attached to each other to form Lichens, many millions of years ago a spirit somehow became attached to the brain of some animal and then both the spirit and the animal profited a lot from this joining together and being together, at the brain surface in attachment.

The animal had a much higher chance of survival because of the guiding intelligence and purposeful movements it got from the spirit, giving its body movements which were intelligent. Natural selection selected for the animals which had the attachment of a spirit to the brain.

Natural selection selected for survival animals which had the symbiotic association of two completely different forms of life, one transparent and completely invisible, free from matter, and the other visible and of solid matter.

The actual start of the symbiotic association in some animal might perhaps have taken place a hundred and fifty million years ago or even before the first dinosaurs existed?

To make it worthwhile the animal would need to have a brain of at least a certain size? Perhaps a brain of about a tenth the size of a chicken brain is enough?

 Technically the attachment of a spirit to the brain was not very difficult since spirits are everywhere. Unborn spirits are not conscious because they are deprived of information.

As soon as they get a continuous steady stream of information they will wake up immediately and become conscious. Becoming conscious means they form images of more than one kind. They can only get such information as they need from brains.

 You could imagine rightly unborn spirits, all transparent and invisible, float about everywhere.

Contact points of an unborn spirit touching matter, could temporarily form everywhere, in the air and even on specks of dust. Though you don't see that.

A vast reservoir of unborn spirits must exist, all of them probably doing nothing but being floating around without consciousness and waiting for so many millions of years to be born, but it's a gamble and only a very tiny fraction of those living spirits will ever be born on Earth. Though you can imagine spirits are floating about everywhere invisibly. The spirits are not floating in our dimension in any form that would take up space, rather they would appear as very tiny minute specks. A single tiny bit of information or data may pass through each speck, which should be called a contact point.

And even small specks of dust in the air may sometimes carry the contact to a spirit.

Spirits are all alive but without having a brain they are deprived of information. And without any information they are unconscious, though perfectly alive as if in a deep sleep.

Now the lucky spirits become attached to the brain of an animal.

By that I mean of course they become attached to an animal before it is born, and while the animal is still an embryo.

As soon as the brain gives them information they wake up and become conscious, and thus spirits are born. But they were absolutely living even a trillion years before they were born, it's just that spirits are invisible, everywhere, you can't see them.

But they are everywhere. There must be a vast enormous perhaps infinite reservoir of unborn spirits which randomly and temporarily make contact with matter at a single tiny point which appears for a moment at random.

All spirits are alive and living, they have eternity to live, even if they are unlucky and never make contact with the brain of an animal, with the brain of an embryo, anywhere on Earth.

They will probably make contact with a brain someday somewhere in the universe even if they have to wait unconscious floating about for billions of years before that happens. They have billions of years or longer, because spirits live forever. Now as you can imagine spirits which are invisible floating about everywhere, they cannot occupy a brain which is already occupied. It's like there is a "no vacancies" sign.

Were it not for that, it would happen often that you would encounter other minds thinking to you. If that ever did happen, don't worry it's harmless. The other spirit has a perfect right to be there, as much of a right to be there as you do. Technically a second spirit might possibly become attached to your brain but something prevents it usually.

How a spirit becomes attached to a brain

In the normal case, exactly how does a spirit become attached to the brain of a tiny little embryo? The tiny embryo of a mammal or of a bird must always find a spirit and then attach that spirit to its brain.

Proteins in brains do all sorts of different things as they work.

I am suggesting that a form of protein is essential in the joining of spirit to a brain. Protein molecules can have so many different properties depending on how they are folded and on their amino acid sequences.

As a floating spirit randomly touches an unoccupied brain cell in the tiny brain of an embryo, be it an embryo of a bird inside an egg, or the embryo of a human inside a mother, the floating spirit actually becomes stuck to joined to, and attached to the brain.

For sure there should be a kind of protein which is specially made to be "sticky like glue" to spirits. There must be quite a lot of it inside the brain. The protein which is sticky like glue to spirits is probably one that has been discovered, only, the scientists have no idea what its real purpose is.

They might have noticed that among other things it can act as an enzyme and then not have wondered about it?

So as a spirit randomly touches the protein inside a brain cell, it becomes stuck to the brain cell as if glue was there.

As it becomes stuck to a brain cell, in the embryo, it must at first make contact at a single contact point.

A contact point has formed between a spirit and that tiny brain. Actually it's more likely that the contact point is formed earlier with an earlier type of embryonic tissue which will turn into a brain and a spinal cord later.

It might be contact with a type of embryonic cell which exists in an early development stage of the neural tube.

The neural tube of an embryo is at first a nearly flat strip of cells which a few days later on rolls up into a tube, and the tube forms the brain and the spinal cord.

As it rolls up into a tube, one end becomes larger and becomes the brain.

Probably a spirit makes its first contact with a cell in that early nearly flat neural tube, actually before a brain begins to form.

From that stage on, contact points where a spirit makes contact with cells, have to divide and multiply.

A spirit is a living thing so its contact can divide and multiply. And those contact points are quite invisible, but they are joined to proteins inside the brain cell.

In an embryo, these proteins must come out of the living cells, diffuse around by diffusion, and float about in the liquid which bathes the neural tube. They might also travel from one cell to another through cell to cell junctions.

These proteins each carry a contact point where the same spirit is stuck to them and as they come out of living cells, diffuse through liquid and float around through the liquid, they spread, and they enter other living cells.

And so the contact between an invisible living spirit and the brain of a tiny embryo increases and spreads.

And probably it takes only a few days, after that starts for that single original contact point to have divided and multiplied and become millions of contact points in the liquid which is in the neural tube, and in millions of cells which will soon form a brain.

And perhaps contact points might exist in the spinal cord. The living spirit is then well attached to the brain.

It becomes conscious when the brain gives it information.

Considering how the spirit naturally needs information in its way of life, and likes information a lot, it may well itself try to increase its contact with the brain after it has received interesting information from it. There could be mainly a passive mechanism, or the spirit itself could try to increase its contact?

And so symbiosis between the two completely different forms of life is formed in the brain, reminding you of a Lichen in which a fungus and an algae live together in symbiosis. A brain is certainly like a Lichen.

A Theory of Consciousness

An important experiment. How to attach a spirit to a brain.

 The most important experiment in the world is to attach a spirit to a tiny little embryo deliberately. Using a small amount of manipulation under a low power microscope you can do it.

What if you experimentally take some brain cells from an animal, or take brain cells from a human brain, and then inject these brain cells into an embryo which is at the right stage of development?

 The right stage of development is probably the stage when the neural tube is just forming, or when it is flat but soon to roll up into a tube.

 The spirit which is already firmly attached to the brain cells which you inject, may come out of the brain cells as contact points. The brain cells injected may release contact points which then spread out to diffuse around the embryo.

So a spirit which is already joined in that normal symbiosis with an older human brain, might then become attached, joined as well to the little brain of a small developing embryo.

A spirit will then have two brains, a spirit having two brains instead of just one. That's very easily possible.

It will create conscious visual images of the objects which are in front of both pairs of eyes, it will see with its eyesight with two pairs of eyes! The process of taking a few brain cell and then injecting them into another small growing embryo, could then be repeated again and again, as many times as you like, giving the one spirit more and more brains.

And as the spirit has more and more brains it will see consciously through more and more different pairs of eyes.

The lucky spirit will through that experiment become attached permanently to more and more different brains, and it will actually be living healthily in all of those different places at the same moment in time. And it will move around several different animals with its will, at the same time.

A Theory of Consciousness

Obviously this would be extremely useful. Imagine that you take some brain cells, a piece smaller than a grain of rice, from a blind man, and inject them into the tiny embryo of a bird, a bird such as a parrot?

When the embryo hatches into a parrot, the blind man will see perfectly with the eyes of the parrot and so he will no longer be blind.

With some practice he will surely learn to read a book by looking at the words with the eyes of the bird!

 The man would fly around making the parrot fly where he wants with his will, and see the world perfectly well with its eyes. This obviously allows you to give eyesight to blind people everywhere.

Will some practice the man would learn to speak with the mouth of the parrot.

He could even hold writing pen with the parrot's beak, and with a lot of difficult practice, he could become skilled at writing with the pen held in his parrot's beak.

I suppose witnesses who see this would believe wrongly that there must be some electronic circuits and a camera hidden inside the bird, and thus find an electronic explanation for the miracle.

But I think I would take a metal detector, and with the metal detector prove that there are no electric wires in the parrot. Obviously in time it would be understood that something quite different is happening, a miracle of a spirit being transplanted.

In a much more advanced experiment, exactly the same thing can be done with a human embryo in a mother's womb, allowing someone to become reborn as a baby, and to be both a baby and an older person both at the same time. It is possible to live in several different places at the same time, and to be both a little baby and a much older person simultaneously.

 If the successful experiment is repeated every generation you will get eternal life. Eternal life is certainly possible, spirits last forever.

So there is a way to harness this important natural phenomena and make it do something slightly different and better, than what it actually does naturally in nature.

The artificial harnessing of this natural effect can bring a spirit to eternal life, as spirits are made with eternity anyhow.

About Eyesight

As the embryo develops and an animal or a bird is born, its spirit gets large amounts of information from the visual cortex area of the brain, which is connected through the optic nerves to eyes. The spirit gets it from its attachment to presumably all of the brain cells inside the visual cortex of the brain.

 Every single time any brain cell in the visual cortex fires an electrical impulse, the spirit receives a single tiny bit of information at a time from that brain cell. When millions of brain cells fire, the spirit gets millions of bits of information. And the information has been organized and properly presented by the brain cells in a way which makes it much more easy for it to be transmitted to a spirit.

And one of the issues is that a spirit may need intelligent clues to allow it to find out the X,Y coordinate positions of pixels within the whole picture. And this need for clues is because a spirit has different physical dimensions which could cause problems in finding X and Y coordinates. Since the physical dimensions of spirits are presumably not the X,Y,Z dimensions of our space.

Many years ago, scientific experiments on the brains of cats proved that in the visual cortex area there are many columns of brain cells, in which individual cells can be sensitive to lines of contrast. They put drawings in front of the cat's eyes, and put very fine electrodes into the cat's brain which could detect individual brain cells firing.

 They demonstrated that individual cells in every column, can detect when a line of contrast has a specific angle. Or detect when a line of contrast has both a specific angle and is moving in a specific direction. This organization in the brain might have several purposes, and I think that among other things, this might give the attached spirit essential clues to help it work out the approximate X,Y coordinates of pixels relative to the whole picture.

And the attached spirit has to learn what the information means. It learns after a while to create a really bright real image of what is in front of the animal's eyes, the whole world existing outside and around the animal, and this image seems to be the real world existing for miles and miles around the animal.

Eyesight creates a whole brightly lit world which is obviously enormous, and it might be sometimes literally mistaken for the creation of the world. When people do not realize that they are seeing an image, they must mistake the conscious visual image for the existence of the whole world. And animals probably always make that mistake.

Obviously such an image is of great value to wild animals and evolution and natural selection, so animals which do attach a spirit to their brain were selected, and survived. Then the animals began to depend on it, and couldn't survive without spirits attached to them.

How a brain controls a spirit, and how the spirit controls the brain.

In any animal brain which has a spirit, there is a continual two way control. The spirit controls the brain and the brain controls the spirit. Both controls each other, and that is essential.

I will explain some details of this two way control.

The spirit controls certain brain cells with its will, causing the brain cells to fire electrical impulses. The best example of that is the motor neurons. Motor neurons have long nerve axons which go down and reach into the spinal cord or even all the way to the muscles.

And as well, the spirit can also stimulate with its will other brain cells whose long nerve axons reach to and control motor neurons.

Most of the motor neurons would be always quiet and relaxed, without a spirit having a direct action upon them. The spirit is attached to the motor neurons with its contact points.

With its will and its wanting to do something, the unseen spirit causes a change in those contact points, and as the contact points change, the motor neuron amplifies the change a million times using enzymes, until maybe in a hundredth of a second the signal becomes strong enough to cause the cell to fire an electrical impulse.

Some types of brain cells are specialized to detect and amplify the change in contact points that are caused by the will of the spirit, and it's rather like the cells in your eyes being specialized to detect light.

In this other case, the brain cells are specialized to amplify the will of a spirit probably through a series of enzymes reactions which carry out amplification.

So as a spirit learns to form the big bright three dimensional image of all the world around the eyes of an animal, the spirit has a will and it wants things.

And as it wants things it learns to control the muscles in the animal's body, to move the animal in a way that gets what it wants. And it does that directly stimulating brain cells to make them fire electrical impulses.

For the learning process the spirit needs a way to store memory. It stores the information of what it is learning in the brain, since a spirit by itself has no memory it has to rely on the brain as a way of storing and retrieving memories.

So millions of brain cells which would be completely quiet and calm on their own are stimulated by a spirit.

So that's one example of a spirit controlling an animal, but there is always a two way control.

The brain has to control the spirit. How does the brain do it?

There has to be a two way control always and continually, the brain controlling its spirit, and the spirit controlling the brain, both ways, and continually. What the brain controls is the will of the spirit.

There is an instinct making system in every animal brain made from circuits of nerve cells. And its purpose is to cause instincts by controlling the feelings and the emotions of the spirit, controlling what the spirit wants and what it tries to achieve.

The instinct making system of brain cells, controls the will of the spirit and controls what the spirit wants and what it tries to achieve.

Immediately the spirit then tries to achieve something, and to achieve something it wants it learns to directly stimulate motor neurons and other brain cells, to contract the animal's muscles in the coordinated way and make the animal move in the direction it wants.

The animal moves and does something purposeful under control of the spirit.

The instinct making system therefore controls what the spirit wants and what the spirit tries to achieve.

And the spirit then controls the body of an animal by its stimulating the motor neurons, and causing a purposeful movement of the animal.

This is the two way control in which each in the symbiosis controls each other. An animal has to have instincts because in the wilds intelligence without knowledge is not enough.

Even if the spirit has the highest intelligence, the animal it leads could die off because the spirit did not actually have knowledge of the dangerous predator which might want to eat it.

Also if the instinct making system did not cause the spirit to feel the conscious feeling of strong thirst at the right time when the animal needs water, the spirit would not actually help to guide the animal towards drinking water.

Hundreds of different examples must exist.. Hundreds of different examples of the two way control actually working must exist.

The instinct making system in the brain of an animal, is made from complex circuits of nerve cells, and it is shaped by natural selection. And this instinct making system controls the emotions and the feelings of spirits, and controls the will of the spirit.

The instinct making system causes the spirit to want things, and the spirit then controls the body of the animal making it do purposeful actions to try to get what it feels it wants.

 The ability of the instinct making system to control the feelings and the will of a spirit, probably depends on a specialized scientific property of spirits, a property of the spirits themselves.

For spirits have existed through eternity attaching themselves to the brains of animals.

And living in that form of symbiotic attachment to the animals, it would be useful to a spirit if its attachment to an animal was easily selected for by natural selection. Without that spirits might have little chance of ever becoming conscious, which they have to do by having firstly attached themselves to a brain.

And so you might think or theorize that if in eternity spirits go through any form of natural selection and evolution, or if they go through any form of natural selection effects themselves, in the very long length of eternity, then it would be a big advantage to the spirits if the spirits have physical properties that make them useful to animals. Something which would make spirits useful to animals includes of course properties which make it easy and successful for the instinct making system of an animal brain to control the spirit's feelings and will.

Roughly stated, I think the scientific property of spirits in this case is that when certain patterns of information are transmitted to a spirit this causes the spirit to feel a conscious feeling or an emotion.

Whenever a simple pattern of millions of bits of information is transmitted to a spirit, the spirit will instantly feel a strong conscious feeling. It happens so long as the pattern of information is of one of the right types. A conscious feeling or an emotion is something which always controls the will of a spirit.

It's the creation of a feeling or an emotion which immediately controls the will of a spirit, and immediately changes what the spirit is going to try to achieve by moving the animal.

And so the instinct making system in the brain of an animal generates patterns of electrical impulses in millions of brain cells, and as these large patterns are sent to the spirit through its physical contact with every brain cell, this immediately causes a spirit to feel either a conscious feeling or an emotion.

Any emotion can be produced in the spirit by a pattern of information generated by brain cells, and these patterns are hard wired by nerve cells and controlled by genetics and by natural selection.

Some examples.

A certain pattern of information generated by millions of brain cells will immediately make a spirit feel fear, the spirit will immediately feel afraid, that is the conscious emotion of fear. And so the spirit will make the animal run, or look around, try to see if there is a danger, and then the spirit will make the animal run away from danger. That helps the animal avoid predators.

A certain pattern of information generated by millions of brain cells, (in the instinct making system) will immediately make the spirit feel the conscious feeling of thirst.

When receiving the pattern of information the spirit feels thirsty, and the spirit will then walk the animal towards a river so it can drink. The spirit controls the muscles of the animal with its will while feeling thirsty, to make it walk towards water.

And then the spirit helps a lot to make it drink.

Since the conscious emotion of feeling thirsty is a desire of spirits to get water. And this feeling always causes spirits to want to go nearer to the water.

So obviously certain nerve cells in the brain must detect physical effects of a lack of water whenever the animal's body is becoming dehydrated, and needs water. And these nerve cells turn on or trigger a pattern generating circuit in the instinct making system, which immediately transmits a pattern of information to the spirit which will cause the spirit to feel thirsty. A feeling of thirst is by definition a feeling that makes a spirit want to go near water.

And as the spirit feels very thirsty, it always steers the animal to get water.

Another example. A certain pattern of information, generated by brain cells part of the instinct making system, will make a spirit feel the conscious feeling of pain. The feeling of pain is something all spirits want to avoid very much.

So the spirit will make the animal run, or look around and if it can see a cause of the pain the spirit will make the animal run away from it, with its control of the motor neurons which make muscles contract.

Obviously all these things are a very high advantage for the survival of a wild animal, and natural selection will select for it!

The instinct making system must contain two kinds of circuits of nerve cells. The one I have already mentioned generates a pattern of information which causes instinctive feelings and emotions in spirits. These circuits need to be triggered to turn on at the correct moment.

A Theory of Consciousness

To turn on the pattern generating circuits only at the right moments, there usually has to be pattern recognition circuits, which trigger them to turn on.

These pattern recognition areas are circuits of nerve cells.

 And what patterns they recognize is inherited through DNA, and is selected for with natural selection. They automatically recognize meaningful patterns of information coming from a wild animal's senses. From its eyes, from is ears, from its taste.

And as they recognize specific patterns from the senses, the pattern recognition circuits trigger activity in the normally quiet pattern generating circuits. Different patterns which can cause different feelings are generated, and they are transmitted to the spirits.

A pattern of information is of course a very different thing from a feeling or an emotion, but what I am saying is that pure spirits have a scientific property which makes them naturally convert certain patterns of information into something completely different, the moment they receive it, and that is a feeling or emotion.

 As an example of it, suppose that a certain wild animal needs to be scared of snakes. Its brain will have evolved through natural selection a pattern recognition circuit which recognizes snakes and becomes active whenever a snake is nearby.

Its inputs may be cells in the visual cortex and the audio areas of the brain, recognizing either what a snake looks like or its hissing sound. When this pattern recognition circuit detects that a snake is very near, it will trigger a pattern generating circuit which generates the specific pattern of information that a spirit converts into the conscious emotion of fear.

That way the spirit learns immediately to feel fear, that makes the spirit scared of snakes, and the spirit will control the muscles of the animal to make it run to get away from the snake.

A Theory of Consciousness

Throughout millions of years a long series of pattern recognition circuits must have evolved which are all inherited by DNA, and each of them is connected though nerve axons to stimulate a pattern generating circuit which will transmit a pattern of electrical impulses to the spirit that is attached to the brain in the symbiosis.

The pattern generating circuits are as well all inherited through DNA. And the pattern which it transmits to the spirit immediately causes a conscious emotional feeling, which may be fear, thirst, hunger, love, anger, pain, hunger, or almost anything.

It has to be the right feeling in a particular situation to give the spirit the desire to carry out the right action.

This must have started before the age of dinosaurs, and obviously there was something else working before the symbiosis with spirits started. All these animals already had instinctive behavior which was not in any way connected with spirits, but which must have been at first just reflexes.

The reflexes worked at an early stage of evolution when brains did not yet ever have any spirit attached to them.

And when the system of attaching and using a spirit first evolved, the animals probably never lost the simpler unconscious reflex system, and this secondary system must remain in animals as a backup system. That is if for any reason the attached spirit does get the right feelings, the older unconscious reflex system may function as a backup.

 And so if scientists ever do a detailed examination of how brains cause instinctive behavior, they will certainly find both different systems working side by side.

Starting to form conscious images, and having to find X,Y coordinates.

How or why does a newborn spirit start to form conscious images.

I mean the images which are very bright and illuminated, and which seem solid. And which of course seem to be the whole world illuminated brightly and existing around you. A newborn spirit has been floating around for infinities, or for trillions of years perhaps, without any form of consciousness because it was not receiving information from a brain as it had no brain.

Only the lucky few, a very small fraction of these spirits will ever in millions of years become joined in the symbiotic relationship with the brain of an animal.

Those lucky few are in nature attached to the brain of an embryo through at first just a single contact point.

You could imagine it as a little round point. The contact points multiply and then a few weeks later the whole brain of a small new-born animal has the contact points where the lucky spirit is attached to most of the brain cells.

When a bird hatches from the egg, or even before that, the spirit gets a lot of new information from the small brain.

A lot of that information coming from the animal's eyes and its visual cortex which will arrange that information in a pattern that is much more easy to understand.

A lot of a brain is designed to arrange information in a form which is more easy for a spirit to receive and interpret correctly.

A spirit might consider raw information from the eyes, as something it can't understand, or as something it does not really receive.

For example, what if a spirit received directly the information that is in the optic nerves, from the eyes, but before that information got to the brain?

Because the spirit has different scientific dimensions from matter, it might not be able to understand the X and Y coordinate system of pixels inside the optic nerves. And it might not be able to tell if pixels are close together or far apart in the image. In other words it might not be able to actually see the image, because an X and Y coordinate system can't be simply transmitted to a spirit.

The spirit would then need plentiful intelligent clues which would help it to work out where pixels in the eye's pictures are in relation to the whole pictures.

It has been discovered long ago, by using electrodes in the brain of cats, that the visual cortex area of the brains of cats is made up of columns and that in each column certain brain cells detect automatically lines of contrast which are at different specific angles. These brain cells fire when there is a line of contrast with a specific angle only.

And other brain cells in the columns are sensitive to moving lines of contrast, and they fire only when the contrast lines are moving in a certain direction and have a certain angle.

And so this information might give the spirit clues about the X and Y coordinates of pixels within a picture, and allow the spirit to receive the picture which it could not receive directly from the optic nerves.

That might be the main purpose of these columns of brain cells, which detect lines of contrast at specific angles. But not the only purpose, since those columns of brain cells would have other purposes as well, including the fact that they would be used by the instinct making system.

So since it has different physics from matter, a spirit might not be able to easily feel X and Y coordinates in our space, and so it might need information clues to give it an ability to find the X and Y coordinates of pixels in an image. Brain cells which recognize lines of contrast at different specific angles, and movements of these lines of contrast, might give spirits those clues, about X,Y coordinates of pixels.

In general information has to be analyzed and presented in the right way by the circuits of brain cells, and then after that a spirit will be able to understand what the information means.

The new born spirit is then actually like a scientist, always trying to work out what the information might mean, always trying to work out what the information is about, and like scientists the newborn spirit instinctively tries to create images and models which will show the subject or what the information is really about.

The unborn spirit is transparent, invisible, and as it is born, it has naturally a scientific mind.

With its scientific mind, the newly born but invisible spirit of the small bird hatching from its egg, or a young animal, tries to understand the information it is receiving scientifically.

And immediately it will start to try to form an image which will show what the world looks like, and maybe what it has so far managed to understand. And then it becomes extremely skilled at forming good images after a learning process.

I have to emphasize that a newly born spirit of a small bird hatching from its egg has got a scientific mind. That is something you're not normally aware of and it works at a subconscious level. With that scientific mind the spirit automatically tries to understand large amounts of information and to form better and better images that show brightly and on a completely conscious level the world that exists outside and around the little bird.

The conscious image of light.

So much information comes from the animal's visual cortex. The spirit that lives in contact with the brain surely tries to understand at first what light is, and as it does not know very much you might think that its attempts to understand what light is and then form an image would not work all that well.

But the image it creates is a suitable one and it is very good. Considering that it does an amazingly good job, and creates the conscious image of light whenever the animal's eyes are in daylight.

To you your conscious image of light is very bright, it is what seems to be an absolutely physical bright light in the world outside you.

And when you look at sunlight or at any bright electric light bulb what seems to be an absolutely physical real bright light is actually in reality your conscious image of light.

 (Just look at a brightly colored electric light bulb and say "isn't my conscious image of colored light bright and beautiful").

 So your spirit is doing a marvelous job of creating the bright image which seems to actually be light, so perfectly like your idea of light you'd never know the difference.

As an animal is born the newborn spirit attached to its brain in symbiosis, naturally has a scientific mind immediately, and as the animal obtains information from its sense of touch with objects, its spirit learns to create images which are more models than images because these images feel absolutely solid!

There is in a spirit a subconscious side. You are not aware of your spirit trying to process raw information, you only see an image which is a final result.

A spirit gets its memory from the brain it is attached to

You need to remember that spirits must have no memory of their own. Without a brain a spirit would forget everything and if it lost its brain it would have amnesia instantly, though of course spirits remain living forever they can do that without having a memory.

I assume spirits can remember absolutely nothing on their own, and so all of their memories depend on their being able to store data in a brain. As well as ordinary conscious memory, there must be a different form of memory that you are not consciously aware of, as it is used technically for the process of creating images. When an animal is born it is most important to the animal's survival that the attached spirit should learn quickly how to form good images of the world, and so a specialized form of brain cell must exist in order to help a spirit do the specialized work of converting raw information into conscious images. In order to help a spirit learn and remember how to form better images, a special type of memory brain cell must exist, in the brain of the young newborn animals.

 So every new born wild animal's spirit has a scientific mind, and gradually over a few days or a few weeks, the newborn spirit starts to turn the raw information into images which are more like models than images, as many or some of them are quite solid.

As the animal grows a little older it may seem to be quite low in intelligence.

What has happened, is probably that a special type of brain cell, required to store information about the specific subject of how to form images, has hardened.

Those brain cells have finished learning, and become more or less fixed. This has lowered the spirit's intelligence a lot.

So the adult animal would have the low intelligence in reality, while when it's younger its lack of knowledge would make it impossible for it to take any form of an intelligence test though it has got the potential for an amazing intelligence because spirits are a more advanced form of life.

The beginning of a spirit's attachment to its brain.

How is a spirit attached to a brain? I believe the contact has to be at very tiny spots called contact points, and they can divide and multiply and that happens when the contact is increasing.

One side of every contact point is a few molecules of protein, and the other side of every contact point is a part of the spirit attached to it.

A comparison to a piece of cloth on a table.

An interesting comparison is made if you imagine that the flat surface of a table is the human brain, and imagine that a piece of cloth or a sweater placed gently on the table is the spirit.

You notice that the sweater placed on the table is raised above the flat surface because it has a third dimension.

It also makes contact with the table at thousands of very tiny spots which are contact points, and they might be seen clearer if the table has been sprayed lightly with tiny drops of adhesive.

A sweater is above the table because it has a third dimensions, and you imagine that the surface of a table has only two dimensions.

A spirit has dimensions which matter does not have.

So a spirit is certainly lifted up to be above the brain, just like a sweater which is resting on a table. Also we are used to saying that the sweater "is on the table," but the sweater is "not inside the table".

That is a very important point: The sweater is not inside the table, instead you say it is on the table.

In a very clearly similar way a spirit is "not inside" a brain, partly because like the sweater at least an extra dimension lifts it up.

The tiny little points at which a spirit makes contact with a brain, are possible even if the spirit has no dimensions at all that are in common with three dimensional matter, and in that particular case I think they would have to be tiny round dots.

About being in many different places at the same time.

A physical property of spirits is important. And that is, a spirit can be in many different places at the same moment of time. The contact points are examples of such places.

And for sure spirits should be never at all detectable in the air around the brain, since it will only remain in our dimensions only at its very tiny microscopic contact points. For sure it's completely false to imagine making contact with a spirit near a brain, because the spirit will have no contact with anything except for the tiny contact points inside brain cells.

So if you take some brain cells out of the brain and move them to some place far away, the spirit should still remain firmly in contact with those separated brain cells, but not anywhere in the air in between the cell sample and the brain.

The contact of the spirit with an isolated sample of brain cells can be detected and can be proved by the right experiments.

And there is basically two quite separate forms of experiments which you can do. I said in one sense a spirit attached to a brain is like a piece of cloth on a table. But in another way, a model comparing a cloth on a table to a spirit, is not quite right. And this next explanation is very strange but important.

Imagine what will happen if a tiny piece of a living human brain is taken out during surgery (It could be smaller than a grain of rice).

The spirit of the man is in contact with all those a few thousand brain cells, and as the small amount of brain cells are taken out, it will remain attached to them and will not under any circumstances have to lose contact, if those brain cells remain alive.

(I suppose that if the brain cells that were taken out of the brain died, then after a couple of hours or a few days, but not immediately, the spirit would lose contact.)

The small sample of brain cells could be taken from a man's brain if he is having surgery for a brain tumor for example.

And this small sample could be placed in a tissue culture solution, or in serum, and oxygen bubbled through the solution to keep them alive and healthy.

For this experiment it's rather important that the small sample of brain cells should be in good condition.

Imagine that the tiny sample of brain cells is put on an airplane and is flown thousands of miles away.

The extraordinary thing is that the spirit of the man is going to stay attached to those brain cells when they are taken by air a thousand miles away.

It won't in any way whatsoever tend to lose contact, and contact won't weaken. Not even by a millionth, and that has a lot to do with a fact that spirits must have physical dimensions which are important to them but which are not at all related to our usual three dimensions, our usual world of three dimensional space is not the world of spirits.

They have a different world. And you might consider that most of a spirit is floating in that different world. It touches our world but is floating somewhere else.

And as a spirit is in contact with some brain cells it will not be affected, not the tiniest tiniest bit by an increasing distance between those brain cells, and the brain they were taken from.

So as the small sample of brain cells were taken from a man, who had brain surgery for a brain tumor perhaps, are taken out of the operating room and flown by air to another country, that man's spirit will be fully attached to that small sample, it will be fully attached to those brain cells.

And distance as we measure distance on Earth will have no effect however slight, however small on his spirit's attachment to the sample, and on its ability to detect information coming from those brain cells.

So if you got an experiment working when the sample of brain cells is kept nearby, the same experiment should work equally when the sample is far away from the man.

And I would be certain that even if you put those brain cells on a spaceship and flew them to planet Mars or to planet Pluto, that would have no effect, not the smallest, not the tiniest, not the tiniest infinitesimal effect on the spirit's being fully still in contact with them.

But how could you demonstrate that his spirit is attached to them?

It would be somewhat difficult, just slightly hard to do, but there is certainly several ways. If you take those brain cells and scan a pattern of information to them from a computer, then depending on what the pattern is, the man might immediately say he's seen a light or he has felt something.

Especially if those brain cells were taken from his visual cortex, his spirit is used to obtaining visual information from them, and so, scanning the right information onto them from a computer should cause a flash of light.

 And he would just say he has seen a flash of light each time. A successful experiment. And that would work however many thousands of miles away the brain cells are flown.

And it is a surprising fact that spirits can live in many different places at the same time.

Attachment proteins, vital for the Symbiosis

 There must be proteins in the brain which have the special job of attaching the spirit to the brain. Like a sticky substance spirits simply stick to it.

 Over millions of years the attachment of a spirit to the brain was improved so that the animals gained more and more advantages from that attachment. Natural selection kept on improving the attachment and the symbiosis between those two completely different forms of life.

Now what forms a conscious image? A spirit exists, your spirit exists attached to your brain. I use the word "attached". That is something which I think could grow into an exact science, I think an exact science will say more about how it is "attached" to a brain.

A Theory of Consciousness

It is attached physically to all of the brain cells all over your brain, all over your whole brain's surface.

As your eyes work millions of bits of information go through your optic nerves, and go to your brain, and from its physical contact your spirit is receiving these enormous amounts of information.

A spirit has its own very high intelligence.

It's not the brain that is intelligent. A brain is not very intelligent.

But a spirit can use a brain for doing things like storing information, and even doing calculations. Unlike the brain a spirit has no memory of its own, without the brain a spirit has no memory.

So the brain lacks intelligence and cannot think, while the spirit has a high intelligence of its own and does all the thinking, but has no memory of its own and on its own cannot remember anything.

Imagine that a spirit were suddenly separated from its brain, the spirit would instantly have amnesia. It cannot remember anything without its brain, and in the symbiosis each gives the other something it needs.

The wild animals gain intelligence in their movements, they gain much better eyesight, and better ability to find food, better ability to escape from predators, better ability to find drinking water, all from the attachment of a spirit to their brains.

There is a symbiosis in which two completely different living things are joined together.

A spirit is a form of life which is transparent so you cannot see it, and it has by itself a very high intelligence. But it has no memory, and to remember anything a spirit must store the information in a brain, and then retrieve the information from the brain, and while the spirit is storing and then retrieving information from a brain the spirit is thinking.

Every tiny spot where a spirit is joined or attached to a brain in symbiosis, should be called a contact point.

Through each of these tiny spots a single tiny bit of information can flow at a time. It is rather like with binary when there is a single tiny bit of information at a time, So when any brain cell fires its tiny electric current, a single tiny bit of information flows at a time to the spirit.

How does a spirit become attached to the brain of an animal?

When the animal was a tiny embryo, inside an egg, or inside its mother's womb, it had at the very earliest stage no spirit.

And then a few days later millions of its brain cells were attached to a spirit.

One spirit became attached to a whole large area, and since spirits are living things it's altogether reasonable to suppose that a part of spirit can divide and multiply. And therefore the tiny spots where the spirit became attached to the embryonic brain were able to divide and multiply and spread out through the embryonic brain.

From at the start just one single contact point, there are a few days later millions of contact points.

And this most likely happens before the tiny embryo has a brain, it should happen in the tissue which begins to roll up and form a neural tube, which then forms both the brain and the spinal cord.

;-------

I once met a certain form of a Ufo air pilot. And he told me not to be afraid, since their officials and their most ancient lawmakers have preserved wildlife in many different planets or districts of space.

And he said that according to their law which would be eternal, the Earth has been given to its wildlife forever, and forever dedicated to wildlife and to a wild and natural habitat.

He told me that he was not trusted to be on Earth since he had once said within hearing of their officials, that he would like to give primitive people eternity. And their officials never forgave him for having said that.

And he said that the only reason why he could communicate with me at all, was that he had just paid an enormous sum of money as a bribe to a corrupt official.

He told me that their corrupt official was one of their most ancient lawmakers, too proud of his legislation to ever allow changes to the law. He said that the law for protecting primitive wildlife was incredibly complicated. And he said that if he spoke to me in a way that would change any form of a planet in any way, that would be most strictly forbidden.

As changes and knowledge introduced of any sort would be like introducing mutations to wildlife, where wildlife was protected.

;------

Experiments with tissue culture and adapted LCD screens

In order to experiment with transmitting information to a spirit, there is a really important experiment which I want to describe. It is made with some equipment which could be made easily in the future.

 Brain cells could be taken from a human being, and spread out on a flat sheet of something which would let you transmit information to them, at any specified X,Y coordinates perhaps, in a layer just a few cells thick.

One could use something made from the same technology that goes to make a laptop or computer LCD screen.

In the LCD screens, there is firstly an address decoder which takes

 an address from a computer, and translates that address into activating a single horizontal line plus a single vertical line.

Where the two lines cross there is a Pixel, which you see as a tiny dot of color. There might be 1 million pixels.

There is one pixel for every crossing of the horizontal and vertical lines, and each pixel is addressed by a computer address.

Typical LCD screens, have some form of transparent electrical conductor to make the X and Y lines transparent, and they have some form of transparent thin film transistors, for each pixel.

The transistors which are there at each pixel, detect when a crossing horizontal and vertical line both have a signal, and then create the voltage which stimulates a liquid crystal pixel to change its color.

What if a single cell thick layer of living brain cells were spread out over this screen? And an oxygenated tissue culture solution is spread out over the screen to keep those brain cells alive and healthy.

What if instead of changing the color of a liquid crystal pixel, those transistors at the crossing points stimulate a living brain cell.

There would be a layer protecting the thin film transistors at every pixel from water, but leaving exposed to the water some very tiny

electric contacts where each pixel would be.

The computer LCD screen is a very common equipment, and it should be modified, keeping the same basic type of thin film transistors, but maybe removing the LCD or liquid crystal layer. And substituting for the colored pixels a slight change to the thin film transistors which are there at every single pixel. The thin film transistors at every pixel, should now be used to transmit very small electric shock stimulus to living brain cells which must be spread out thinly over the screen.

Now with your computer program you can specify the X,Y coordinates of the brain cells to be shocked or stimulated. And if you wanted to you could try simply scanning pictures onto it, from for instance bmp images or a computer video.

Now if you imagine that the living brain cells were taken from a living person, such as a man who needed surgery to remove a brain tumor, what would happen if you attach this equipment to a computer and scan information onto it!

The man would immediately feel something, he would know when the computer is on off, and he would say "I have felt something".

If the brain cells came from the visual cortex area of his brain, the man might very well say he is seeing flashes of light!

Yes if he said he is seeing flashes of light that would be a very successful experiment.

Endless experiments can be done with this equipment.

The man would say he felt something or has seen flashes of light whenever the computer scans the data to the screen with his brain cells.

But when the information sent to it is of a form which is meaningless then he might feel nothing, and it's a question to investigate, what patterns and what arrangements of information are meaningful to a spirit? Now imagine the computer uses a little camera and that you scan the picture from the camera onto the screen with his brain cells alive there in tissue culture solution, then perhaps after a while he would say he sees the picture.

But, as in the visual cortex of a brain information from your eyes has to be analyzed a lot by brain cells, (for instance there are columns of cells for at first recognizing lines of contrast of different angles, and recognizing lines of contrast which move in different specific directions) it might be essential in the experiment to make the computer analyze the information from the camera in a way that exactly imitates the visual cortex of a brain, before scanning the information onto the screen which has his live brain cells.

A computer program could imitate a visual cortex, and do what that does.

There would always be a learning process, as the man might at first see extremely poorly. But after months of practice his spirit would sometimes gradually improve what its forming images, and in time produce good conscious images.

In the end if the computer is properly programmed to imitate what happens in a visual cortex of a brain, the man would certainly see with the camera as if they were his new eyes.

And it would be quite extraordinary that he would be able to see with it wherever it is. In fact his spirit would begin to form bright conscious visual images, which would be like the creation of a new world in the room that has the camera. Obviously he would be in two different places at the same time.

Always a spirit can be in many different places at the same time, but only in places where the tiny contact points that join a spirit to brain cells exist.

A working computer simulation of a brain's natural instinct making system

This is about the equipment in which something like a modified LCD screen would have brain cells spread out over it, and when it is connected to a computer the pixels would directly stimulate brain cells which touch them. Several other quite different experiments can also be done with the same equipment.

It is possible to use it to find out how the instinct making system in the brain of wild animals works, not by using any wild animals at all, but using just one human volunteer.

If you tried scanning many different kinds of patterns of information, just by luck certain kinds of patterns of information could make the man who donated a few of his brain cells immediately feel a conscious feeling.

It could give him immediately unusual feelings. It could do anything, such as happy, sad, angry, frightened, thirsty, hungry, itchy, contented, almost anything. Because spirits probably have a scientific physical property which makes them more easily useful to wild animals, and this is that when certain particular kinds of patterns of information are transmitted to them, it causes them to immediately feel conscious emotions and feelings.

A spirit translates a pattern of information into something quite different which is a conscious emotion or feeling. And almost any kind of feeling can be caused by different patterns.

The brains of wild animals obviously have an instinct making system, and this includes an ability to generate many different patterns of information, and transmit the information to a spirit using millions of brain cells. The system uses that property of spirits quite a lot.

And this is necessary for survival and constantly used in the brains of wild animals, to cause feelings or emotions which control what the spirit wants, and what it tries to achieve, and so causes behavior.